Perspectives of Nature
Volume 3

Perspectives of Nature Volume 3
ISBN: 978-1-945307-35-5

Copyright © 2021 by Paul Košir
First published April, 2021

With the exception of brief quotes for the purpose of review, no part of this book may be reproduced or utilized in any form or by any means, electronic or mechanical, without express written permission from the publisher.

Published by
Nature Works Publishing
1235 Denton Street
La Crosse, WI 54601
sciromanticpoetry@gmail.com

Book designed by Rodney Schroeter

Perspectives of Nature

Scientifically Romantic
and
Experiential Nature Poetry

Volume 3

ACKNOWLEDGEMENTS

I must gratefully acknowledge
Rodney Schroeter,
whose confidence from the beginning
in my scientifically romantic style of poetry
made this project imaginable
and whose advice during the process
made the project achievable.

I also must gratefully acknowledge
the contributions to this work
of the members of the
La Crosse Area Writers Group,
and LAWG Poets,
who helped me to polish my good poems
into publishable works.

Finally, I must gratefully acknowledge
my wife, Lilly,
for her cheerful and helpful responses
to my frequent questions and complaints
while dealing ineffectively
with computer situations.

Author's Introduction

For more than two decades, I wrote no poems.
This followed my initial burst of nearly a dozen poems beginning in the late 1980s, when I wrote poetry that described the science behind natural phenomena and processes. It was picturesque, instructive, and sometimes light hearted. Early in 2012, I emerged from my dormancy with "Sun Dogs," a poem that appears in my first collection, *Perspectives of Nature*. I wrote it to see if I still was adept at writing poetry after time off as a father, teacher, naturalist, and historian. It seemed I still had the knack and that I also had begun to develop my own style.

   I continued to write poetry in this style and early in 2020 published my second book of poetry, *Perspectives of Nature, Volume 2*. That volume had fewer meteorological and astronomical poems and more "story" poems. This trend is also found in *Perspectives of Nature, Volume 3*.
The poems in the third volume, like those in first two, are what I call "scientifically romantic," not because they accurately describe the natural world without romanticizing it (which is the case), but because they encourage personal feeling, a trademark of the Romantic Movement in poetry.
William Wordsworth, a founder of the Romantic Movement, said "poetry is the spontaneous overflow of powerful feelings."
May you have many 'Wordsworthian' moments as you read the poetry in *Perspectives of Nature, Volume 3*.

   Many of the poems in this booklet use scientific terms. Definitions of more obscure terms and less-frequently-used words can be found on the left-hand page opposite the poem in which the words or terms occur.

   Readers may want to use the blank portions of the left-hand pages for journal entries, recording when and where events described in the poems are observed. Locations of sandstone outcrops, going on a bird count, visiting a bog or prairie, and finding young animals might be recorded. Listening to crickets, observing bats or water striders, and seeing monarch adults, caterpillars, or eggs all are worthy entries.

Writing in this booklet would not ruin its leaves, but cause them to flower.

                                                       --- P.K.
                                          March, 2021

To Lilly,

who's always been supportive wife
in my not-always-normal life.

Table of Contents

Author's Introduction v
Time 1
Nature's Clocks 3
A Useful Force 5
Heat 7
Light 9
Sound 11
Three Worlds 13
Swimminng 15
Flying 17
Wind 19
Plate Tectonics 21
Geysers 23
Sandstone Beauty 25
Soil 27
  Lichens 29
  Prairie 31
  Marsh.... 33
Bogs.............................. 35
Oceans 37
Water Striders 39
Crickets 41
Monarchs 43
Bats 45
Woodpeckers 47
Woodpecker Species 49
A Tale of Two Sparrows 51
Flower Adaptations 53
A Hunt 55
Young 57
First Bird Walk 59
The Towhees of Turkey Hollow... 61
Incredible Edibles 63
The Count 65
Television 67
Mountains 69
Zoo 71
The Grammar of Nature 73
Rime Ice..........................74
Hoar Frost........................ 75
Index............................. 77
Also by the Author................ 78
About the Author................. 79

Perspectives of Nature Volume 3

TIME

The thread of Life rolls out in Spring
to near infinity,

and ties together living things
in Time's divinity.

Our Summer tablet holds a list, a simple registry
of tasks and fancies for our lives that we in mind foresee.

We turn the pages, ledger days against imagined Time,
our Summer stolen by a count not done by final chime.

At Autumn's feast on Winter's eve,
Time cools in hibernation

while other patrons quickly eat,
Time warms in their migration.

How long, it seems, the Winter lasts
as marked by Winter Wren,

count seven seconds for each song
that's absent from his glen.

Notes

Perspectives of Nature Volume 3

NATURE'S CLOCKS

Radioactive rocks are placed in special hourglass,
thence change away at even rate to see the ages pass.
These elements take certain time to change one half away,
same length of Time, new half will change while other half will stay.
In using such an hourglass to time the age of rock,
one finds the dates of nearby stones, as sure as if by clock.
The older rocks are lower down, the younger higher still,
Time stacks things not from top to base, no matter what our will.

Uranium is often used to find the age of rocks,
but diff'rent elements show well the lesser ticks and tocks.
For objects made in human times, the dating that works best
is carbon for organic things, a basket or a vest.
Most items found from ancient times, were made to show respect,
yet objects of prehistory, survival did reflect.

Notes

Perspectives of Nature Volume 3

A USEFUL FORCE

It keeps the Earth in orbit 'round the Sun
with day-and-night rotation being spun
at perfect distance for the Life on Earth
to give its outer layer living girth.
On surface, it keeps boulders in their place,
in elevation of the earthen face.
Environments of humans are the frame
that, held by forces, will remain the same.
When we, like early humans, leave our caves
by simple trail or footpath that one paves,
if in an auto or on back of horse,
our transportation needs this useful force.
With varied movements made from here to there,
perceptible, at best, or on a tear,
machines in use a hundred times a day,
a wheel, a rope, and brushing dirt away,
for harvesting and cooking all our food,
by force, we eat and drink what we have brewed;
with force, the words that humans ever spake,
on paper, now, or digital we make;
More than just opposition, friction drives
the most important actions of our lives.

Notes

Caloric
> From the middle to the end of the 19th century, the fluid that transferred heat according to the obsolete "caloric theory."

Kinetic
> Related to or resulting from motion.

𝓟erspectives of 𝒩ature 𝒱olume 3

HEAT

Before the nineteenth century, "caloric" carried heat
in scientific papers and while walking down the street.

This liquid explanation of how warm things got warm
became replaced by energy, kinetic in its form.

Kinetic heat is passed along in one way out of three:
conduction, radiation, and convection we can see.

Conduction must by contact spread from warmer to the cool,
warm things, when felt, start losing heat; this state can sometimes fool.

Convective heat by currents moves, in rooms and Gulf and core.
Heat rises to the ceiling, so it's cooler near the floor.

The Gulf Stream carries water to the European states,
which makes them warm for folks to live and betters, then, their fates.

Currents in its outer core cause Earth's magnetic field
that saves us from Sun's cosmic rays, without it, fate is sealed.

By radiation, heat is passed to objects far away,
thus ultraviolet energy is carried by Sun's ray

then through transparent barriers, including greenhouse glass,
and CO_2 or methane or some other greenhouse gas.

These U-V rays are turned to heat that's trapped below the sky
to warm the Earth and make some think disaster's coming nigh.

Notes

Planar
    In a plane, flat.

Obliquities (*oh-BLICK-wi-tees*)
    Irregularities in a planar surface.

Convex
    For lenses, thicker in the center.

Concave
    For lenses, thinner in the center.

Angle of incidence
    Angle at which light enters a mirror, lens, or prism.

Newton = Sir Isaac Newton (1642-1727)
    English physicist and mathematician, who taught at Cambridge University.

Whit
    A particle of the least possible amount.

Perspectives of Nature Volume 3

LIGHT

To the looking glass and back, describes the path of light,
reflected from its silvered rear, that's polished clean and bright.
The image seen is regular, transposed from left to right.
What else has turned to wonderland in image of our sight?

By moving to another glass, we change reflected view,
the mirror now not planar, so its image is askew,
which alters our perspective so it's something wholly new,
removing some obliquities that quite unnoticed grew.

But through the looking lens and on, will cause a bended ray.
An arrow 'hind a water glass will point the other way.
The shape of lens, convex, concave, decides how it will stray
from angle of its incidence, where it will never stay.

With mirrors and a lens inside, he built a viewing tool
to help his students learn at night in most impressive school.
As Newton worked, invented math, new physics, as a rule,
he looked at planets' orbits while he sat upon on his stool.

Through the prism and beyond, is where white light is split,
revealing spectrum locked within, six colors to be lit.
The rainbow's seen at other times, in water spray, oil bit.
Light's nature is duality, part wavy and part whit.

Notes

Perspectives of Nature Volume 3

SOUND

Sound travels through a medium, most oftentimes the air,
which, when disturbed, moves molecules to make them dense then rare.

Compressions and releases cause the frequency, or pitch,
the volume, and the quality that makes a noise sound rich,

the beauty heard from instruments or choir or a bird,
the melodies we used to sing or other music heard.

But music's not the only good in sounds that we all share,
communication, spoken, sung, shows others that we care.

All animals communicate. What sets us, then, apart?
Ideas that are new to us and poetry that's art.

Notes

Escher = Maurits Cornelis (M.C.) Escher, (1898-1972)
  Dutch graphic artist.

Perspectives of Nature Volume 3

THREE WORLDS

Through Escher's eyes we Nature see
when we at lake count worlds at three.

With one perspective for each view,
each lake in front of us is new,

for closer to the surface seems
fish image on bent sunlight beams

and tree leaves that on surface rest
look true on every wavelet crest.

The 3-D trees before our eyes
transform to 2-dimension guise

and stretch out over water's face
to Nature and Her tree-line trace.

Three worlds are found in Nature's light,
beyond, below, and water-height.

The images in mind we make
are archetypes for every lake.

Notes

Dallol (*DAL-awl*)
    Region of hyper-saline acid geo-thermal pools in Ethiopia, one of which appears to support *no* life forms of *any* kind.

Perspectives of Nature Volume 3

SWIMMING

The ducks that dabble swim on top, the divers down below.
With bones that are not hollow, though, still deeper can loons go.

A coot will dive then pop back up, but grebes are birds that sink.
They force out the unwanted air and make their air sacs shrink.

Auks dive for fish and feed on squid, but mostly they eat krill.
Antarctic penguins dive to stay away from 'whales' that kill.

Whales with beaks and elephant seals dive deepest of them all
with packed-in blood cells, shunted flow, they make a complex call.

The most-adapted animals to life beneath the wet,
can spend it most completely in an underwater set.

In water, fish do everything, including breathing air,
extracting from it oxygen, dissolved and rather rare.

With their gills attached to arches, which look like slotted 'cheeks,'
most fish use countercurrent flow for oxygen from creeks.

In Dallol, adaptations fail in habitats most cruel;
none live in hyper-saline acid geo-thermal pool.

Notes

Secondaries
   The inner flight feathers that run along a bird's 'arm.'

Primaries
   Elongated outer flight feathers, the 'fingertip feathers.'

FLYING

Aerodynamic shape of wings
helps birds stay in the air

while soaring or in powered flight,
wings working as a pair.

The secondaries of a bird
give wings their shape and lift,

while thrust comes from the primaries
to make birds' flying swift.

The way in which birds ply their wings
Determines path of flight

How much, how hard their pinions move
affects the flying height.

Bald eagles can do barrel rolls,
Ospreys dive feet first,

Hummers aim for nectar tubes
To quench their nagging thirst.

Kingfishers plunge from branch for fish
Crows 'row' their wings to fly

Male woodcocks zig-zag up and up
then tumble from the sky.

Notes

Cyclone
   A weather system in which winds rotate inwardly in a counterclockwise direction (in the Northern Hemisphere).

Perspectives of Nature Volume 3

WIND

The 'magic' that we see on Earth,
                                         to move without a touch,

is done with pressure gradients,
                                         on one side there is much.

For higher pressure moves the air
                                         to where the pressure's light,

which causes breezes from the sea,
                                         but from the land at night.

When cyclones march across our land,
                                         which takes about a week,

these pressure systems aim the winds,
                                         if normal, strong, or meek

High pressure systems, clockwise out.
                                         Low, counterclockwise in.

Our winds, in general, westerly
                                         without an added spin.

The howling winds in Winter chill
                                         then calm at night and freeze.

The gusts of March raise kites, balloons,
                                         while dancing papers tease.

The gently moving summer air
                                         evaporates to please.

  With autumn currents, pollen spread,
                                         to make some people
                                                           sneeze.

Notes

Subductively
: By means of subduction, in which the denser tectonic plate moves under the one that is less dense.

Harry Hess (1906-1969)
: American geologist who first advanced the theory of seafloor spreading.

Alfred Wegener (1880-1930)
: German geologist who put forth the idea of continental drift, a contentious theory that was not accepted until after Wegener's death.

Perspectives of Nature Volume 3

PLATE TECTONICS

A suit of armor plates the Earth with pieces of the crust
that fit together puzzle-like, into their places thrust
by hotter mantle down below that rises up and slides
so plates tectonic pass on edge, build tension 'long the sides.
Plates also move subductively, too slow for us to see,
as lighter continental crust slides over crust of sea,
yet smashing plates will mountains build, if density the same.
There's spreading in Atlantic Ridge was Harry Hess's claim.
Volcanic action, earthquakes, too, are found near boundary,
where plates colliding drive the course of core geology.
The movement of tectonic plates, or Continental Drift,
explained by Alfred Wegener, caused scientific rift.

Notes

GEYSERS

Like a tempest in a teapot,
a geyser makes its show

if two outlets reach the surface,
both heated from below

by magma underneath the Earth
which makes the water grow.

When hot enough, the water boils,
steam rises in long spout

till it releases in a jet
and all the steam is out.

The water, just erupted forth,
fills up the 'room' and then

is heated up another time,
boils over once again.

A geyser's faithful to its blasts,
you'll always know just when.

Notes

The Ordovician Sea
    Sea that covered parts of North America
    from 485 million years ago until 444 million years ago.

Perspectives of Nature Volume 3

SANDSTONE BEAUTY

Four hundred million years ago,
the Ordovician Sea

lashed the shoreline, pummeled sand,
without impunity.

Through millions more, sand turned to stone
with iron showing hints

of many oxides colorful,
red, yellow, rusty tints.

These shades are hid in many spots,
but still considered rare,

in glens and dells and overhangs,
wherever sandstone's bare.

In Winter and in early Spring,
the overhangs are best,

endowed with frozen waterfalls,
beginning at the crest.

The frozen jewelry worn by Earth
has beauty seasonal,

but deeper colors of the Earth
are less ephemeral.

Notes

Profile
    The arrangement of the constituent horizontal layers of soil.

Horizons
    The various identifiable layers: O, A, E, B, C, and sometimes R (rock) that make up soil.

SOIL

The bedrock of an area is parent rock of soil,
produced by Nature and by Time, no touch of human toil.

Rock broken by the freeze and thaw of water on the land
shaped mineral ingredients, like clay and silt and sand.

By color, texture, structure can a soil be classified;
in profile of horizons, can development be spied.

Material from Life on Earth, organic in its source,
imparts the chemistry of Life, but not the living force.

Such matter and some broken rock are all soil needs to form,
then sculpted by the climate and by weather, sun or storm.

Organic matter, decomposed, is food for plants to grow,
if in a place by Nature set, or planted in a row

Eroded soil is gone for good, take care of where we 'tread.'
Formation takes a thousand years, so soil is limited.

Protect the soil and use it well, do not the landscape hurt.
Remember that our gift from Time, our soil, is more than dirt.

Notes

Fruticose lichens
    Lichens that look like tiny shrubs.

Perspectives of Nature Volume 3

LICHENS

On the Isle Royale and to its north west,
where looking for lichens is always the best,

with so many lichens of so many kinds,
people go out on hikes and count up their finds.

They find them on trees and bare ground, solid rock
in morning and evening, midday on the clock.

The flat, crustose growth form that some lichens take
produces same surface old spray paint may make.

The foliose lichens appear to have leaves,
while fruticose lichens reach up with long 'sleeves.'

The shape of a lichen comes from fungal part,
with acid it uses to break rocks apart.

This rock-breaking role of the lichens is key
to add to nutrition of soil that is free.

The fungus is shelter for algae to live
and make all the food that it's ready to give.

Their housing-for-food keeps communities strong
in complex arrangement where life forms belong.

Next time you're in Nature and find yourself hikin,'
remember to stop and reflect on each lichen.

Notes

Seres (*sears*)
    Stages in secondary ecological succession, which occurs after an area is invaded by plants.

Perspectives of Nature Volume 3

PRAIRIE

Sedges have edges and rushes are round.
but grasses are hollow from top to the ground.

First two like wetness, but grasses not so;
they're native to Plains states, where winds often blow,

Growing in grasslands called prairies by name;
the rich soils beneath them made landscapes to tame.

Prairies remain on the lands undisturbed,
but only in places where plowing was curbed.

Flames on the prairie, ignited with ease,
burn all the dead grasses and smallest of trees,

Properly managed and set through the years,
they guided succession, in stages, or seres.

Grassland savannahs have trees of great girth,
with deep-growing roots that help anchor the Earth.

Climax of plant life may not be the trees,
but grasses and flowers in warm Summer breeze.

Notes

This poem was inspired by the La Crosse River Marsh,
a 1200-acre wetland in the middle of La Crosse, Wisconsin.

MARSH

A bur-reed marsh is mostly there,
where cat-tail plants are very rare.

While threatened species fly around,
invasive species grow from ground.

Marsh dries each Mississippi flood
to make it Mississippi mud

in the year, Two Thousand One
and '50s, '60s, still not done.

Teachers use for education,
others find their recreation,

A place where scholars do research,
and spirits treat it as a church.

Blackbird, egret, dragonfly,
a wood duck painted on the sky,

fishes, muskrats swimming past,
gosling follows mama last,

swallow, warbler, leopard frog,
turtles sunning on a log.

All these things seen in "The Marsh"
where Summer's gentle, Winter's harsh.

Notes

BOGS

If drainage of a pond or lake is blocked by native clog
for thousand years, a wetland forms, a spongy "quaking bog."

The vegetation that can live within this acid pool
will grow in mats a meter thick, unsteady stepping stool.

A bog is poor in nutrients, most native plants need more,
but not bog's insect-eating plants, which have enough to store.

Decaying matter in a bog is starting point for coal
Absorbent peat moss, found in bogs, provides free flood control.

The peat moss in the garden shops combats aridity.
When grown in bogs, cranberries have a mild acidity.

Amend the garden, keep it moist, red berry juice to sup.
Next time it feels that you're bogged down,
think rather, you're bogged up.

Notes

Gyre (*JAI-er*)
    Enormous region of rotation in an ocean that can collect countless items of debris.

OCEANS

Most never lose the sight of land,
and still we say we love the sea.

Experience is beach of sand,
beyond the surf is absentee.

Our love of sea vicarious,
so what we learn becomes heartfelt.

The oceans' fates precarious,
whenever ice sheets warm and melt.

But even with an airplane view,
can't fathom its immensity.

The circulation 'round, anew,
is caused by water's density.

Cold, salty water, weighing more,
submerges with Atlantic North,

moves downward onto ocean floor
to rise with Gulf Stream flowing forth.

In the ocean named "Pacific,"
where water forms gigantic gyre,

refuse has become horrific
in garbage patch of plastics dire.

Notes

Jacana (*zhah-SAH-nah*)
A member of the Jacana family, wading birds with extremely long toes.

WATER STRIDERS

Jacanas walk on lily-pads
and houseflies walk on walls,

squirrels can climb electric poles
and cave fish, waterfalls.

How simple the relationship
that formed an ancient bond,

imbuing water-striders means
to walk across the pond.

The surface tension water holds,
supports the insect's 'feet,'

adapting its anatomy,
so miracle's no feat.

With wings, the strider's unconfined,
yet stays in habitat

to take advantage of its traits
that work where water's flat.

Notes

Stridulate (*STRID-yu-late*)
Of an insect, rubbing two body parts together to make a noise.

Dynasty of Tang (AD 618 – 907)
Rulers of the Chinese Empire who venerated crickets and kept them in cages.

Perspectives of Nature Volume 3

CRICKETS

When male crickets want to mate,
they rub their wings and stridulate
by scraping 'scraper' 'cross their 'file'
a nearby female to beguile
with buzzy chirps in charming song,
again, again, the whole night long.
Most people find that crickets soothe
but others wish harsh chirping smooth.
Cave crickets, "mute," of wings devoid,
make some with basements feel annoyed.
Crickets chirp from planting rows
till harvest days come to a close
These bookends make the cricket sage
to those who shrine them in a cage.
At times in Western countries, jeered,
throughout the East, they are revered
for all the chirping songs they sang
since time of Dynasty of Tang.
Where crickets dwelt, good fortune found,
but here all live upon the ground.
Inside, a cricket stops its chirps
to signify invading perps.
Outside, crickets get a feel
for temp'ratures they then reveal.

Notes

Instars
  Larval stages of an insect (caterpillar stages of a butterfly).

Chrysalis
  Colorful (usually green) casing that grows around the fifth instar of a butterfly caterpillar (moths make cocoons).

Predaceous(*pree-DAY-shus*)
  Predatory.

Perspectives of Nature Volume 3

MONARCHS

From eggs laid under milkweed leaves, the caterpillars grow,
devouring the bitter leaves by munching off each row.
The bitter taste is glycoside, a poison to the heart,
tolerated by the monarchs, though toxic from the start.
The only job for larval form is eat and eat and eat
For dozen days they do this well and never skip a beat.
Fifth instar of this larval stage will change to butterfly
once it's emerged from chrysalis and after wings are dry.
The western monarchs hibernate, the eastern fly in fall.
They overwinter Mexico, three thousand miles in all.
Five generations fly return, they break it up in parts
and fly a route that's new to them without a map or charts.
'Our' monarchs use two habitats, two places where they dine
The loss of milkweeds here *and* there has led to steep decline.
With climate change, predaceous birds, and maybe hit by cars,
the monarchs are not kings and queens, but butterflying stars.

Notes

Nash = Ogden Nash, (1902-1971)
    American poet who wrote a poem entitled *"The Bat."*

BATS

Like Nash, I rather like the bat,
its high-pitched chirps are never flat.

Too high to hear for humankind,
all bats, not blind, by echoes find

the flying insects that they eat
with wayward skeeters for a treat.

erratic flight to catch the bugs
that veer too near their mouse-like mugs.

With furry bodies, wings of skin,
stretched tight between long fingers thin,

some types of bats migrate in waves,
while most bats hibernate in caves.

The vampire bat is never found
where northern species hang around,

but places south, like Mexico,
and movie screens with vampire show.

Notes

Hyoid bone
    The strong, flexible bone supporting the tongue, the "tongue bone."

Verberation
    The striking of an object or body that causes sound.

Perspectives of Nature Volume 3

WOODPECKERS

Unique among the Aves Class are birds that peck on trees
to find the insects hidden there, uncover them with ease.

With two toes forward, two toes back, they show accepted stance
when posing for the picture books, but need a second glance.
They cling to trees with two toes up, but aiming down, just one;
the fourth toe braces off to side, where tight embrace is won.

The bird, so propped, can drill for food by using bill to peck
with quick, repeated movements at near twenty times a sec.
This rapid-fire pounding's more than any bird alive,
they make ten thousand pecks a day in order to survive.

The wood is not what 'peckers seek, they're feeling for some bugs
to catch with lengthy barb-ed tongues by many little tugs.
Attached to base of nostril right, tongue loops inside the head
and exits in between the bills to keep the 'pecker fed.

Their tongue may act as cushioning so stress of impact low.
Air pockets that are found in skull let not vibrations grow.
Between the eyes, woodpecker's skull and hyoid bone, as well,
are spongy, so absorb most shocks and verberations quell.

Near all of impact energy through body dissipates.
The bit that's left will stay in head where heat it generates.
The bill, as well, does redirect the energy of pecks.
Third eyelids at last-second blink, protecting eyes like specs.

The birds that peck know how to act, no need for any thought.
Our engineers could learn from them, if Nature's ways are taught.

Notes

Perspectives of Nature Volume 3

WOODPECKER SPECIES

Red-headed Woodpeckers
The young wear plumage of adult, but color broken gray.
Storing nut and insect finds, it covers them each day.

Red-bellied Woodpecker
Two times I've seen the blood-red splotch that's always facing tree.
Their odd behavior might be play, but does it give them glee?

Yellow-bellied Sapsucker
Return in Spring, drill holes in line to make a well of sap
for yellow bellies, first to sip, then hummingbirds to lap.

American Three-toed Woodpecker and Black-backed Woodpecker
These species have three toes per foot, two forward, but none back.,
their stiff woodpecker tails that prop make up for toes they lack.

Downy Woodpecker
Nuthatches and some chickadees make winter flocks seem odd.
Not just on trees, a downy pecks on galls of goldenrod.

Hairy Woodpecker
With larger size and longer bill, a hairy needs not wait;
at feeder, downy, hairy meet, the hairy first to sate.

Northern Flicker
It doesn't slam its head on tree, it pokes its bill in ground
and searches for the insect life that's living all around.

Pileated Woodpecker
These crow-sized birds dig feeding holes some inches deep and wide,
a few so very long and bare that insects cannot hide.

Notes

Unfurled
    Spread out (in this case, after unpacking).

Perspectives of Nature Volume 3

A TALE OF TWO SPARROWS

Eighteen fifty was the year a shipment sent from Liverpool
arrived at Brooklyn Institute, a New York center with a school.

Not music, art, nor lit'rature, what did the parcel feature?
A label read, LIVE ANIMALS, so package held a creature.

Packed by hand most gingerly, the birds arrived alive,
the eight pair never really "caught," in spring they "failed to thrive."

A larger shipment ended well, another, even better.
Philly ordered thousand birds, it said so in a letter.

House Sparrows lived in dozen states by eighteen hundred seventy,
when building houses for this bird became the fad of plenty.

Many argued, Get them out! They eat not insects as we thought!
There's no more room for native birds or others that we've brought!

Digesting really starchy grains, adapting well to urban land,
they filled a new and growing niche, their numbers rose, got out of hand.

As habitats too urban turned, their numbers were declining.
In Delhi, people build for them grass houses with a lining.

∞   ∞   ∞   ∞   ∞   ∞   ∞   ∞   ∞

In eighteen hundred seventy, from sparrows of the Olden World,
were twenty or two dozen sent, with feathers still unfurled.

Called German or Eurasian Tree, with insects on their dinner plate,
the birds preferred farms, parks and towns, aggressiveness was not a trait.

It's on one hundred fifty years, they've had no smoother sledding,
Saint Louis never far away, their U.S. range not spreading;

but back in Europe, some decline, while Asian population fine.

Notes

Perspectives of Nature Volume 3

FLOWER ADAPTATIONS

Oh my darling, Oh my darling,
Oh my darling, columbine,

You are red with yellow insides
and the hummers drink your wine.

From the spurs on backs of flowers,
they sip nectar oh so fine.

Pollination is the mission
they accomplish as they dine…

Oh my jewelweed, Oh my jewelweed,
Oh my jewelweed, touch-me-not,

Flowers look like bee garages,
rub off pollen on the spot…

Oh my milkweed, Oh my milkweed,
Oh my milkweed, monarch food,

Bees step into flower buckets,
to their feet is pollen glued…

Oh my orchids, Oh my orchids,
Oh my orchids, they take time,

For the sites of reproduction
and the smells you also mime….

Notes

Perspectives of Nature Volume 3

A HUNT

The Past is our remembrances and they become our lore,
a chance to live a second time or more or more and more.

Survival's based on memory and aptness for the kill.
The struggle is primordial, the battle rages still.

Few second chances in a hunt for animals to try.
Who flees the first, if safely done, has one less chance to die.

No winner in the contest while combatants both alive.
No loser if, in battle fierce, contestant does revive.

In all-out war, on either side, there's little left to give.
The celebration at the bell, is more of Life to live.

They shake no hands at end of hunt in game called Life they play.
If conflict draws, they scamper off; each goes a separate way.

Notes

precocial
    independent in many aspects of movement and feeding soon after birth or hatching.

altricial
    helpless in many aspects of movement and feeding for a long time after birth or hatching.

Perspectives of Nature Volume 3

YOUNG

Born or hatched with opened eyes,
precocial young are ready;

they don't need parental help,
first steps and stance unsteady.

Some enter Life with eyes closed tight,
Young altricial need more care,

All things 'round are new to them,
They begin with bodies bare.

Ungulates, the hoof-ed beasts,
in rearing are precocial.

So, too, are hares, yet rabbits, no,
they're very much altricial.

Is the womb where young develop,
As precocial lone or twin?

Or after birth, altricial,
with the rest of litter kin?

Precocial or altricial?
Which describes the human young?

Precocial by their single birth,
yet altricial lullabies sung.

Notes

FIRST BIRD WALK

I poorly planned my nature walk for first day in the park,
we watched the birds along the bluff, but hike soon left the mark.

Each step became more treacherous, no longer on the trail,
my leadership had showed us naught, but Nature did not fail.

Some dripping water showered birds, one green with yellow breast,
the other bird less camouflaged, with striking colors blest.

The red and black of male distinct, their contrast weakens knees,
but female held to different scale, her subtler colors please.

The male began the toileting, alighting rather low,
adjusting his positioning to dip beneath the flow.

Clear water splashed upon his head to make his body clean,
he perched upon a nearby twig and thereupon to preen.

The female scarlet tanager, at basin after mate,
used her time to cleanse her skin and comb her feathers straight.

Each sat before the fountain drips and took a final turn
then followed Nature, formed a bond, eloped behind a fern.

Notes

Perspectives of Nature Volume 3

THE TOWHEES OF TURKEY HOLLOW

I met them all one day in May,
on hikes we'd call on each.
From trees in grasslands they would sing,
their songs like human speech
The first we'd hear was "Normal Ned,"
with "drink your tea" from tree.
Next visit was to Simple Steve,
who only said, "drink tea."
Two pathways led to "Double Don,"
who sang in double notes.
"Drink-drink tea-tea" is what he said
in almost double quotes.
With "Backward Bernie," words got turned,
he always made us think,
if his songs were admonishment,
how do we "your tea drink?"

Notes

Perspectives of Nature Volume 3

INCREDIBLE EDIBLES

One year while working at the park, I did a cooking show.
The weekly program was a hit, once filled up every row.

With forty to one hundred guests, I had to move with haste
to serve each one in audience a little sip or taste.

The food I served I cooked that night or in preceding week.
Before the meal we'd take a hike, more edibles to seek.

One night a man at table last, refused to eat the nettles.
He'd seen all other groupings eat the nettles dish from kettles.

He'd even watched his family eat and say to him, Great, Dad.
But still refused… Then ate a dab. He grudged to say, Not bad.

Another night, when show was done, a couple came, was glad.
When you said nettles had the taste that spinach also had,
we worried that our son would balk, for spinach was his 'pain.'
He loved the nettles best of all, we really can't explain.
I spoke to them, behind my hand, here's your new spinach plan.
Next time there's spinach on his plate, say Nettles from the can.

MENU
Tossed salad of edible greens
Soup: Dandelion broth
Appetizer: Morel mushrooms (from freezer)
Entrée: Nettles Italiano
Dessert: Wild ginger candy or
Berry flummery (by season)
Beverage: Nettle tea

(From *Billy Joe Tatum's Wild Foods Field Guide and Cookbook*)

Notes

Prelate (*PRELL-at*)
    A bishop, abbot or other high church official.

Protonotary (*proh-toh-NOH-tahr-ee*)
    One of the high prelates in Rome.

Prothonotary Warbler (*proh-THAHN-oh-tare-ee*)
    A swamp warbler named after the protonotary.

THE COUNT

Before we paddled park canoes on weekly guided float,
I always talked (or did I boast?) of nesting birds we'd note.

One morn I told two twins we'd count, in case I often lied;
I wished to show my word was good, for me, a source of pride.

The siblings, both fine Catholics, knew all their prelates well,
including Protonotary, though difficult to spell;

Prothonotary Warbler, too, the quarry for our trip;
"We'll listen for its 'zweet, zweet' song," my one and only tip.

By knowing what their voice is like, "We'll count ten," I averred.
The girls were young and liked to learn, especially brand new word.

To let a little tension build, I warned, "We must not fail."
at my request, their count was mute till fully 'round the trail.

Their parents in the next canoe, the daughters were intense;
they wished to show to mom and dad adult intelligence

though bursting with their happy news of birds seen on the route,
"Our number came to twenty-nine!" (There never was a doubt.)

Notes

Perspectives of Nature Volume 3

TELEVISION

When Dad got home to dimming sky, he changed from clothes he'd worn.
We watched our shows, my sis and I, cartoons were in the morn.

Dad worked, with never being seen, he snuck behind the house.
While glued to television screen, we were quiet as a mouse.

Dad stole back in and hung his hat, then kissed us on our heads.
When shows were done, Mom gave a pat, we climbed into our beds.

Next morn, Mom pried us from the set, she heard me rant and rave.
Outside stood Dad, his coat still wet, in drift he'd dug a cave.

The snow of cave trapped bits of air, so transferred little heat
to insulate our little lair and make it be a treat.

We played all day, were never cold, protected from the wind.
Our TV plans we put on hold, my father only grinned.

Notes

Perspectives of Nature Volume 3

MOUNTAINS

Our
train was late,
but all was good,
we saw three moose on road,
a cow, two calves that pranced
around; their playful nature showed.
We pitched our tents in mountain
dark and slept in peace of night.
Then found we'd camped near mountaintop.
Surprise in morning light. We later saw the bighorn sheep,
a mountain goat, and bear. The mountains are for those in life
who dare to care and share. Now, those who make the mountains prime,
and also take their time, can find a place to mount a hike to see sights most
sublime.

On
diff'rent trip,
with backpacks tried,
We'd hike up to the pass.
Three times I slid from slanted
Tent and slept on snow, not grass.
Our scouts hiked up toward mountain pass
To find the snow neck-deep. We always knew
our plans might change, aescending mountain steep.

Notes

ZOO

There's never "zoo" heard in the word
that some pronounce "zo-ology."
They rhyme the "zo" with show then know
'bout lives of critters that we see.

To view the life we cannot see,

try micro-(sized) biology.

To learn about a honey bee,

the path is entomology.

For those who catch fish in the sea,

their course is ichthyology.

For those who wish to make some brie,

first they should know mammalogy.

If first two 'o's sound as in "zoo,"

there's not an 'o' for "-ology,"

yet if the first 'o' rhymes with do,

there's plenty for "zo-ology."

But if you want to plant a tree

for Arbor Day dendrology,

close down the zoo and you'll be free

to study plants with botany.

Notes

Subjunctive
In grammar, the mood of a verb that is used to express wishes and contrary-to-fact conditions.

THE GRAMMAR OF NATURE

Nature does not live in the subjunctive,
conditions never contrary to facts,

expressing zero wishes and desires,
responding not with language, but with acts.

Inherent message is imperative,
the unrelenting drive for more to live,

keeps going without question or a doubt,
with objects of the actions still to give.

Considering no possibilities
that predicate outside the present tense,

lacking grammar for temporal thinking,
conceiving of the Future makes no sense.

Life is indicated, not suggested,
by Nature's normal ventures, not the rare,

predicting not the movements yet unseen,
She stumbles not on step that isn't there.

(HOAR FROST)

RIME ICE

Nature's silver frosting forms
in winter fog at night,

reflecting from the shrouded moon
a muted glistening light,

which pierces through the veiling cloak
to make the darkness bright.

In stillness, cooling moisture helps
the fragile rime accrue

as midnight's chill the vapors touch
to freeze the morning dew.

From icy wisps of silent air
the frozen prisms grow,

pure water from the misty clouds
shapes crystals into snow

that unseen falls upon the twigs
and feathers boughs with white.

[To maintain scientific accuracy, the poem *Hoar Frost*, on page 3 of *Perspectives of Nature,* has here been renamed *Rime Ice* (above) and replaced by the poem *Hoar Frost* on page 75.]

Rime – frost.

HOAR FROST

On cold, clear nights when soft winds blow
and stars are pinned to sky,

no clouds to blanket frigid Earth,
no snowfalls keep Her dry.

Then coldness wrings the ice from air,
at times from smokestack plumes.

It leaves the icy window scrawls
in drafty sleeping rooms.

In coldness, Mother Nature frosts
all things with crystals clear

by deposition of the ice
from water vapor near.

No liquid water e'er appears
as She prepares the jewels

that decorate the surfaces
left bare till landscape cools.

Deposition – changing from a gas to a solid without ever being a liquid.

Index

A Hunt..............................55
A Tale of Two Sparrows.........51
A Useful Force.....................5
About the Author..................79
Also by the Author................78
Author's Introduction..............v
Bats..................................45
Bogs 35
Crickets 41
First Bird Walk.....................59
Flower Adaptations................53
Flying................................17
Geysers.............................23
Heat...................................7
Incredible Edibles............... 63
Lichens 29
Light..................................9
Marsh.......................... 33
Monarchs..........................43
Mountains..........................69
Nature's Clocks....................3
Oceans.............................37
Plate Tectonics 21s
Prairie...............................31
Sandstone Beauty..................25
Soil............................. 27
Sound...............................11
Swimming..........................15
Television..........................67
The Count..........................65
The Grammar of Nature............73
The Towhees of Turkey Hollow...61
Three Worlds......................13
Time..................................1
Water Striders.....................39
Wind................................19
Woodpecker Species................49
Woodpeckers......................47
Young...............................57
Zoo..................................71

Also by this author:

Perspectives of Nature

and

Perspectives of Nature
Volume 2

About the Author

The scientifically romantic nature poetry of Paul Košir has its academic roots in his nine years as a student at the University of Wisconsin-Madison. There he earned bachelor's degrees in math, natural science, and history. In 2010 he received his Master of Science in Natural Resources and Environmental Education from UW-Stevens Point.

The experiential poetry was drawn mostly from his twelve years as the naturalist at Wyalusing State Park near Prairie du Chien. He also drew on this background to write articles for *Wisconsin Natural Resources* and *La Crosse Magazine* and to publish the book, Wyalusing History.

Košir has taught biology, physical science, and math at the high school level and earth science, biology, and environmental issues at the college level. As a naturalist, he taught all ages about nature through hikes, programs, and displays, something he still does occasionally.

Born in Milwaukee, Košir now lives in La Crosse with his wife and their two sons. He enjoys writing, hiking, bird-watching, gardening, traveling, and visiting relatives.

Photo credits:

Front cover – the poet's wife on Cascade Pass Trail, taken by the poet in North Cascades National Park.

Back cover – the poet's wife on Cascade Pass Trail, taken by the poet in North Cascades National Park.